SCADA

...you must know it before the first project.

MyScadaWorld

SCADA WinCC HMI Profinet Siemens Wonderware InTouch OPC FactoryTalk Ethernet Modbus

MyScada

Copyright

ISBN: 9798630670373
Imprint: Independently published

MyScada

Introducion

I love to create visualization systems. Every time we meet with a client and present the technique suggested by my team I feel great. Automation and 6 strict superior functions have been involved in nearly ten years. Many times I meet people who do not quite understand what process automation, scada, HMI etc. is a few weeks ago. I decided to gather basic information - I hope that I hate theory in an accessible form - and write this short book. Technical but understandable to people who are unrelated to the subject. Before you start searching for answers on the Internet in various forums, use the search engine and use the following. I have collected some basic information for you.

Many times I have encountered a situation where the client does not fully understand what SCADA is. He imagined it as a collection screen that controls local work. It was difficult to prove that the system offered needed strong computers, servers or very expensive licenses for a number of addresses. Collecting data and relying on my experience wanted to show in a concise way what SCADA is but what it is not. What functions does it have, available or how to recognize an advanced system.

MyScada

Questions about what is scada?
How should I build the use of the system?
Is my factory supposed to be modern industry 4.0?
Which system should you choose?
Do I need scada, hmi?
Which supplier should you choose?
Many of my clients ask these and many other questions
when they modernize lines, create new devices, etc.
Can you be someone looking for something new and in
the near future I would like to be a SCADA programmer?
Remember that the topic is very extensive and I could
write hundreds of pages about the details and approaches
of some elements. But do you need it? Is it not better to
take these items as knowledge pills thanks to which you
will learn the whole?
I invite you to read.
Enter the world of SCADA and remember:

You can't repair the world with just one SCADA!

What's SCADA

One of the basic questions you should face is what SCADA really is? If we consider Wikipedia information, **SCADA is an acronym for Supervisory Control and Data Acquisition.** Generally speaking, it is a program running on a computer, communicating with PLCs installed in the field.

The system is additionally built of components such as:
- Hardware
    - Computers
    - PLCs/PACs/RTUs
    - Communication modules
    - IT infrastructure

- Communication technology
  - Ethernet (profinet etc.)
  - RS232
  - RS485
  - Modbus
  - MQTT
  - OPC
  - DNP3
  - IEC 61850
  - IEC 60870

- Software
  - Operator interfaces
  - Engineering interfaces
  - Communication software

SCADA has many tasks and functions, some of which must be fulfilled and some of them are additional. Generally we can separate SCADA functions:
- Display and operate with system views (operator or operators can decide on activities throughout the entire production line from one place)
- Display status and animations of motors, valves, sensors

- Communicate and operating with PLC and all things connected to it e.g. motors, valves, sensors
- Data logging of process values (e.g. temperature, pressures, positions)
- Display trends (printing, exporting)
- Alarm logging (alarm, failure, error, event)
- Display alarms (printing, exporting)
- Control processes in local or remote localizations (some SCADA can me worldwide)
- Allow operator to see malfunction in real-time
- Reporting and printing specific data
- Connection and operate with databases
- Possible to integrate with MES and ERP
- Long term history (Historian)
- Integration with high level languages (C#, .NET, JAVA, VBS)

MyScada

SCADA structure

Meeting with scada created in the 90s of the last century and the visible ones we can now wonder how to divide it. Below I will present a way of division by structure or individual functions.

Structural:

- Monolit scada - computer connected directly to PLC that controls objects.
- Networked - computer connected to local network and communicate to PLCs
- Distributed - distributed system of computers. Possibility to exchange data on server-client structure. It must have been 20 years until the IOT world (now).
- IOT SCADA - modern way everything connected to everything. Very powerful but not easy to complete because it's still not possible 100% in older companies. It's really nice in new factories.

MyScada

On the other hand, we can divide SCADA systems for functional reasons. Each of the following has its advantages and disadvantages. I think it's worth knowing what a given type of application offers us and what variations you can expect in them.

Stand alone

This configuration is the basic mode and is a simple connection of the station to the control system. This configuration is used in small installations where one operator station supervises the operation of the entire system. The advantage of such a system is the centralization of operations and low implementation cost due to the low infrastructure requirements. Such a system is also not difficult when it comes to maintaining operation and maintenance. Thinking about such a system, we can imagine a computer that is connected to one (or several) devices and performs its SCADA functions. Sounds simple right?

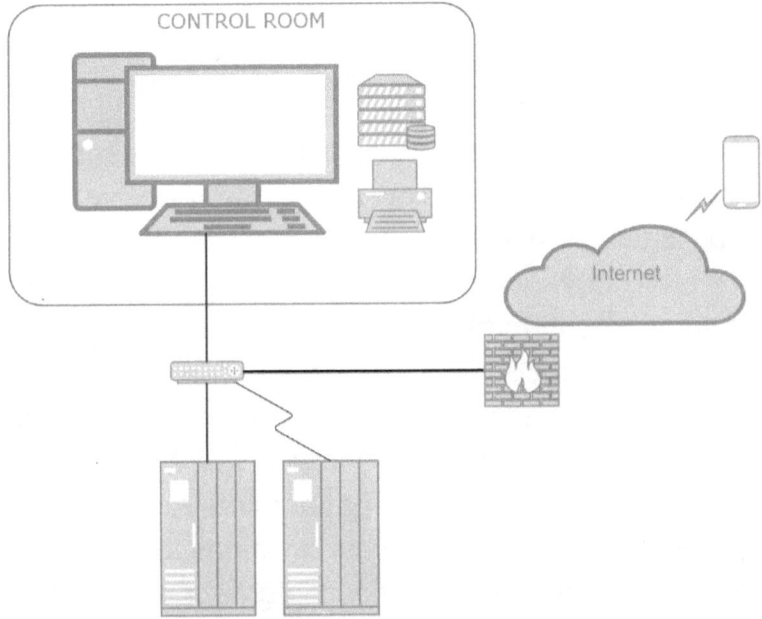

The above graphic shows a stand alone system.

Server - client

The more extensive version is the server-client option. In
this case, I have a server running the main parent
application to which one or more clients are connected.
This allows you to manage multiple operator stations using
a superior automation system. The server supports up to
32 connected clients, providing them with process data,
archive data and messages. Clients can be connected
through a local network or through web browsers. This
structure requires a much better infrastructure on the
production site side and when it comes to web access also
a well thought out and divided internet network. The
advantage of such systems is the ability to be controlled
by many users at one time. You can build an application
so that one customer does not see the activities of the
other and that each customer can work independently.
The disadvantage of the system is primarily a problem with
finding the right implementation company, because a
poorly constructed system in such a structure brings more
problems than a stand alone project.

MyScada

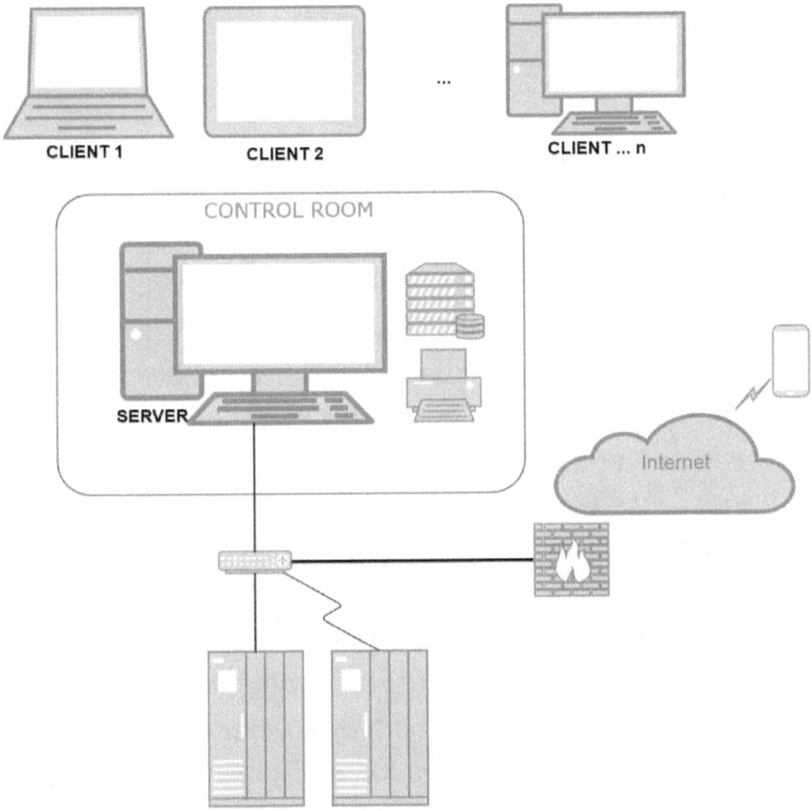

CLIENT 1

CLIENT 2

...

CLIENT ... n

CONTROL ROOM

SERVER

Internet

The graphic above shows the diagram for the server-client structure (1 server and 5 clients)

Redundant servers

The most technically advanced solution that occurs in the structures of created scada systems is the option of redundant servers. It is usually a server-client system with an additional redundancy package. The redundancy option is based on the MASTER server and SLAVE server being in the network. Both check constantly whether everything works correctly, both collect historical data, etc. When the main server crashes to which clients are connected, the latter takes over all its functions, clients connect automatically to the second data source and from the operators' point of view there is no downtime . When, for example, a server that has broken down is repaired and turned on again in the network, both servers synchronize data from the time when the lack occurred.
Such a system is created for 24/7 production where you cannot afford downtime associated with disk, computer, etc. failure.

MyScada

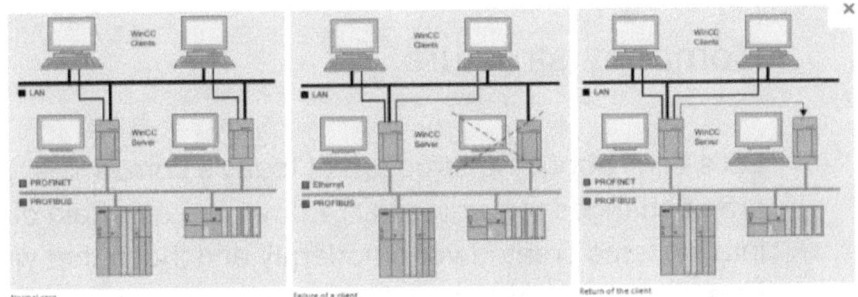

Fot. Siemens.com

What's redundancy?

Company structure

Let's think about the structure of today's companies. In the era of changes regarding IIOT4.0 we see a certain division into systems created very modernly and those that were created, e.g. 5 years ago. Here are the main features of both cases:

Company structure before Industry 4.0	Company structure in Industry 4.0
Structure determined by hardware	Flexible machines and structure
Connection only between production levels	Communications between all participants
Products not integrated into process chain	Products are participants
Roles and features tied to hardware	Roles are distributed in whole network
Databases on demand (local or server rooms)	Big data and cloud computing
No real time improvement	Products can communicate each other - all time

MyScada

	product progress - real time improvements (e.g. Tesla, BMW)
Weak possibilities to preventive maintenance	Preventive maintenance

ERP - Enterprise Resource Planning - generally, it is the entire enterprise management layer where only coarse data and statistics are needed. We know how many products were produced at what time, what was their cost, etc.

MES - Manufacturing Execution Systems are computerized systems used in manufacturing, to track and document the transformation of raw materials to finished goods. The MES system is designed to transfer data from SCADA systems to ERP on an ongoing basis.

SCADA - Supervisory Control and Data Acquisition

HMI - Human Machine Interface

PLC - Programmable Logic Controller

SCADA vs HMI

We often meet people who ask "What is the difference between SCADA and HMI"? We must explain to a non-technical person why he must pay X more for a product that is the same for him.

I would start by explaining what HMI is. Well, HMI is a human machine interface or an element that allows "communication" with the device. It is generally an application running on an operator panel installed in a control cabinet. Such an application performs the functions of:

- communication with PLC
- process visualization
- alarm display
- collecting and displaying process data (trends)
- gives the ability to control individual elements such as valves, engines.

When we read the above, we see a number of similarities to SCADA.

MyScada

They differ in the scale of operation. The larger the scale of control, the greater the chance for SCADA.

HMI - small scale
- local control of a part of the production line
- usually one machine or just limited operator functions
- Usually HMI is a part of SCADA

SCADA - small, medium and large scale
- Centralized or distributed control system
- controlling the entire production line, e.g. 20 PLC and 20 HMI connected in a system,
- controlling the entire production hall

The market approach is that additional processing is often needed, which is why SCADA systems are more common. In this case, the HMI is used as an integral part of the system, e.g. adapted to the local work of a fragment of a single line or a single functionality (e.g. loading station, where the operator can confirm the batch loading in accordance with the production list).

Of course, the system based on operator panels is cheaper and allows the implementation of local machine control. The problem arises when we would like to

MyScada

combine our control and management system to be able to obtain global remote control with monitoring of historical parameters. In particular, when we think about the future possibility of implementing a production management system, MES class systems. In this situation, the choice should be obvious. The implementation of the SCADA system - even individual applications - will allow us later - in an easy and cheaper way to implement the main IT system, which is difficult to achieve using only operator panels.

Panel manufacturers claim that the operator panels have advanced archiving and remote access capabilities, but I know from experience how difficult it is to extract archives. Data archiving is usually carried out either in a closed - binary format (which must be converted later, e.g. to Excel), or in text files or in csv format, which are saved on an external memory card or USB stick.

They generally do not have hard drives, only memory cards. This solution also forces the need to trip and synchronize the archive with historical files manually, which is not always remembered in production applications. When the memory becomes full or damaged, further archiving will become temporarily unavailable, which irreversibly results in the loss of process data. In

MyScada

some production facilities it's a dangerous and unacceptable situation because of human health.

Operator panels also have their advantages and are irreplaceable from an economic point of view, in particular in small, autonomous applications that do not require hardware resources or have limited functionality of data archiving.

Below in the table I have prepared a comparison of SCADA and HMI systems in terms of the most important parameters.

	SCADA	HMI
Application size	Scalable from low to distributed on high range	Low (in most cases 1 HMI to 1 machine unit)
Used hardware	PC computer, servers and virtual environment	Panel
Screen	Internal as touch screen, external and multi screens	Generally internal touch screen
Operating system	Generally MS Windows	Own or Windows CE
License	License demand on tags number (e.g. 2000, 8000, 64000)	Limited to HMI version (some limits to screens, tags, connections or functions)
Recipes	unlimited	limited
Databases	Wide range of solutions (e.g. MS SQL, Oracle)	Generally not used

Connection to 3rd part applications	OPC, VBS, C, .NET, MS Excel	Limited, sometimes by OPC
Communication protocols	There is near no limit for protocols. Sometimes additional software / konwerter is needed.	Ethernet - standard, Profibus, RS as option
Server - client	yes	no

MyScada

How to choose the right SCADA?

In the case when we need to manage production, monitor parameters, quality factors, archive data in a continuous and available online, carry out an automatic audit of parameter changes, monitor the work and performance of employees or perform production reports - we have to think WHAT SCADA system will be the best and not whether it is necessary.

Choosing the right SCADA system that will meet the assumptions and expectations is not easy, and ignorance of the functionality of individual SCADA systems supplied by various software vendors may lead to the inability to expand the application or purchase additional licenses that would be included in the basic variant with other manufacturers. It may also cause that despite the purchase of additional licenses, we will still not achieve the intended level of information technology management of technological processes.

The times when the SCADA system was understood only as a visualization and process control application are long gone. The hardware capabilities and development of this

type of application meant that visualization is only an addition to a number of other, more important functionalities, necessary or facilitating the management of the machine park. It even happens that the graphics module of the application is premeditated and the system works as a service on the server. The graphic part of the control interface can be replaced with other elements, e.g. operator panel, and other functionalities - only by developing dedicated applications from scratch.

My system has potential for SCADA?

During the design of the automation system, the industrial automation integrator tries to explain to the user why it is sometimes worth paying more at the beginning of use for a SCADA system or a computer-server platform, so that the designed system would be open enough later that even IT solutions could be implemented in the future until now the user did not need or lacked funds for.

It is the employees of the target plant who use the system and they should care about managing the SCADA platform in a continuous, simple, fast and economical way.

Integration companies that design SCADA applications derive the greatest profit from the preparation of applications, individual function modules and integration of the prepared system with the IT system used in the company. The margin of licenses resold to the end user is usually a secondary matter. Therefore, for developers, unlike the user, choosing a specific SCADA system is not important.

The most important thing is for the user to understand what functions the SCADA system offers. The decision on the selection of the necessary functionalities is one of the most important that should be made at the beginning. An integration company should only provide the user with advice and an offer of available solutions.

The same applies to the MES. The concept of production management functionality is very broad and can be interpreted as a single functionality module, but also as a huge IT structure with the servers that support it.

MyScada

How to choose the right system?

How to choose the right system so that it can be integrated with the IT system or a larger ERP system?

Nobody wants to change the ERP system implemented in the enterprise, because huge funds were spent for this purpose, and its change or advanced modification would absorb next ones. That is why it is so important to maintain a specific manufacturer's standard or to change it quickly, if the first choice turned out to be insufficient.

Most often, the choice of the SCADA system is dictated by the introduction in the plant of the standard used by one software supplier or by the offer to choose an automation system integrator. Software standardization is a very good solution, facilitating the integration of this type of application and, more importantly, a service. From the maintenance point of view, there is nothing worse than synchronization of platforms provided by several manufacturers. It often happens that connecting applications and exchanging data between them is difficult to implement or impossible.

MyScada

The unification of SCADA application software can also affect the reduction of license costs, mainly developer licenses (necessary for editing projects), which are required for each platform separately.

10 questions you have to ask yourself

The next step in choosing SCADA application software are the functional elements that seem necessary to the user. The word "seem" is used here deliberately, because usually the end user cannot define all, and the development of this list is often continued after project talks with the integrator.

Below are suggestions for questions that help in the selection of functionality. The answers to them determine the choice of the SCADA system manufacturer.

- Continuous, reliable and efficient archiving of process values and alarms: should the archive values be viewed only in the SCADA application, or should it be open and available to other applications?
- Should the IT system analyze, calculate and display other values, such as power or energy, based on already collected data, prepare time charts and

MyScada

graphs of dependence of many values or groups of devices?

- Should it monitor and display measurements of e.g. utilities, calculate their consumption and demand, calculate the unit costs of consumption per manufactured element or a specific production time?
- Is the system supposed to control the quality, efficiency and working time of operators?
- Should the system automatically audit the changes in parameters needed to achieve quality and performance?
- Is the system to monitor and calculate production factors, such as downtime, breakdowns, availability, machine use, controlling performance standards or comparing the work of different operators?
- Is the system supposed to track the produced batches due to e.g. production time, unit cost, material consumption, type of materials, person making, current values, e.g. during weighing, etc.?
- Will the recipe system be used during production - starting with the simplest parameter sets - as well as production based on pre-prepared lists in the ERP system, planning the implementation of such recipes, queuing, etc.?

Databases

The graphic interface can be designed in every SCADA system so that it looks almost identical. However, in addition to the graphics module, the most frequently used functionality is the continuous and reliable archiving of production data of all kinds, including those described earlier. The fact that the data will be "saved" somewhere and become archived data - is only part of the success. After all, process values can be saved even by the operator panel. Important, however, is their format and the ability to use it in an open, free, any way, even in the future, yet unknown to the user at the time of implementation.

The best way to store data from the SCADA application that meets these criteria is the database system. Well designed, it guarantees certainty of recording, openness to other applications and preparation of many different reports from the same data. Storing data in a database is often a complex problem due to the choice of IT platform. When choosing a database system, we are faced with the choice of the right database, i.e. the manufacturer of database software.

Not every producer of SCADA software allows you to freely save your own data structures, and if it allows, it is associated with a license to create your own archive. So we have a choice between using mechanisms for creating database applications on the terms of the SCADA system manufacturer and purchasing the appropriate licenses, and creating our own database system and mechanisms for accessing databases and connecting it to the SCADA application.

As a rule, an enterprise already has an IT system based on the database software of a particular manufacturer. Why not use it and collect data in the access place of all industrial and office applications? In our history of automation, no user has ever asked us about the selection of database software, and license savings resulting from the use of existing solutions are often greater than the preparation of the application itself.

Unfortunately, very often, when deciding on specific SCADA software, we are forced to buy a license of a specific database manufacturer. If the user does not use this software in the enterprise, he still bears the cost of the license of the base engine, although he will not use it. If at least the database system is of the same manufacturer as used so far in the enterprise - then, despite the licensing

MyScada

costs, we have the opportunity to seamlessly implement the database - reducing the load on the company's data server and ease of synchronization.

Therefore, one should consider the existing SCADA software solutions that do not require a license for the database server and give the opportunity to choose both the database producer and the location of the tables with data (other servers). Although such software exists, unfortunately it has not become the standard for SCADA system manufacturers.

Remember, however, that in the event that the database server is not attached to the SCADA software as an integral part, despite the savings on the licenses of this server, the database system may still be needed.

Then we have to think about:
- do we use database software from the same manufacturer that already exists in the company,
- do we use a reserve of licenses that may already be purchased in the enterprise but are not being used,
- on what operating system platform (e.g. Windows, Linux) do we want to put our production bases,

MyScada

- whether we will use free database server solutions, which are often sufficient for production applications - then we do not incur any additional license costs.

It is also best to take into account the opinion of employees from the IT department, because they know best which resources can be used and will help in making decisions.

Communications protocols

Another important issue when choosing the producer of the SCADA system is the way of communication with the PLCs used on the machines - that is, with the data sources that we want to display and archive in our SCADA / MES system. In the SCADA software market, you can often find systems that have very limited communication capabilities with PLCs from various manufacturers. It may turn out that the user-selected SCADA system will connect to many PLCs, but not all. It is therefore necessary to check which PLCs (which manufacturers) the user already has in the machines and whether the selected SCADA system has built-in communication drivers for all controllers from which we want to read information.

MyScada

OPC - bridge for integrations

The specific communication driver used in SCADA applications is the so-called OPC server and client (OLE for Process Control), which enables communication with other applications, integrating e.g. an Excel workbook directly with the system, which allows creating reports in the Office environment.

If OPC is part of the SCADA system, it can also be a supplier of data from controllers for other visualization applications that cannot be read directly. Such a solution is often used in applications as a communication "bridge" between various SCADA software producers. However, I believe that this solution should be used as a last resort and not as a basic one.

It is worth paying attention to SCADA applications that are hand-written by automation integrators. Their development may be slower than the products of larger companies, but very often such an application is enough to apply specific solutions, and the price is much lower. However, in this situation, the user is "sentenced" to the company through which the application was written, and any service and development may be difficult.

Often, such applications do not have dedicated communication drivers and use the OPC driver, for which - in a commercial case - you also have to pay.

When using a free OPC solution, it may turn out that the communication will be unstable and will not guarantee continuity of work.

Reports - core of integrations

Another difficult issue related to the selection of the SCADA system is the possibility of using the reporting module. Sooner or later, every SCADA system user who has a database for collecting information will need an appropriate data reporting system.

Many years of experience in using commercial applications shows that the function of creating reports based on a module built into the SCADA system does not fully meet the expected requirements, and at most is a printout of current data, e.g. labels.

Many SCADA software manufacturers also allow the purchase of an additional licensed reporting module.

MyScada

Some providers explicitly state that they recommend using an external commercial reporting application. And here again the question arises whether the company uses any commercial application that allows you to create new reports or whether the environment used to create reports will be compatible with the SCADA system.

Integrators often focus on the possibility of preparing open reports that can be edited or created new without additional licenses, and their cost is only associated with programming work, which can also be done independently.

The versatility of reporting functions is the unlimited possibility of their preparation, simultaneous access for many users, displaying in an application independent of visualization, e.g. in a web browser, remote access, etc.

In order to obtain reports that meet the above assumptions and are based without restrictions on logged historical data, it is necessary to ensure the proper construction of the database system, which will enable data sharing in many aspects.

MyScada

Server and virtualization

The last important point in the selection of the SCADA system is the ability to install applications on server platforms, multi-server and multi-client work.

When the SCADA system supports many machines and production lines, it may turn out that the computer on which the application has been installed will be so loaded that it will simply not be able to work properly. That is why it is worth considering investing in server computer platforms instead of desktop servers. Such solutions are of course more expensive, but less reliable due to the removal of the location of servers from the production area. If the computer visualizing the application on production is damaged, we can replace it without additional preparation. However, if the desktop application computer running on production is damaged, it will cause a long production downtime, and the data collected in the system may no longer be recoverable.

Server platforms also allow decentralized functions of the SCADA system, which reduces the unit load of the computer operating the system. In today's IT solutions, server virtualization is increasingly used, which combined

MyScada

with redundancy gives over 99 percent. chances for trouble-free work.

Therefore, when deciding on server platforms, you should use virtualization, which allows you to easily transfer servers to another hardware platform and facilitates backup.

Licensing - half price of investment

When choosing variants of a given software, one should carefully look at the issues of licensing solutions of software producers and "shell out" the costs of all licenses that will be necessary in the operation of the system.

Sometimes it is worth paying more, obtaining more functionality and guaranteed service from the manufacturer, than to use the own applications of implementation companies. It all depends of course on the price and possibilities.

Unfortunately, despite careful calculations and planning to purchase software licenses, it may turn out that the user has not taken into account additional licenses - often significant for the budget. An example is access licenses to the Windows server operating system and MS SQL or

MyScada

Oracle database server (for commercial versions). **Such licenses may cost more than a prepared application**. They are independent of the SCADA software, and the integrator does not have to provide them to settle their part of the application preparation contract. Unfortunately, we know from experience that server licenses are rarely audited. They should be the responsibility of the IT department, although they often do not want to have anything to do with the production part.

The user's lack of awareness about the need to have a license does not justify him and causes that the law is violated by him and not by the integrator.

Final decision

To choose the right SCADA software provider, please note in which operating system we can install this application. The selection of the database server with which the system will work is very important. It is worth examining whether it is possible to use your own, existing hardware, software and license resources. It is necessary to prepare an appropriate production database structure that will be the source of reports and confirm that selected report writing technologies will be able to cooperate with our SCADA system. That is why it is so important for the user

to be aware of what he expects from the SCADA system and the integration company.

When the user, thanks to his own involvement, reaches the assumed functionality of a system prepared specifically for work in his enterprise, he will have simple access to the generated information. The user can expand the system over time, as well as postpone the costs associated with this expansion. Otherwise, he will pay for the system, which he will not be able to fully use, and the analysis of the excess information generated by the system will be completely unprofitable in relation to the time lost.

TOP 3 SCADA distributors

My choice - powerful, flexible and verified SCADA

Below is a (subjective) list of the most interesting SCADA systems available on the market. I have known and used all three systems personally for many years and will gladly present my feelings regarding each of them. When it comes to the global market of individual systems (Siemens WinCC, Wonderware InTouch, Rockwell FactoryTalk) it coincides with which markets which systems need.

From my observations it appears that just a few years ago in Europe ruled Siemens and in America and China wonderware. Rather, Rockwell would be a local North American product. The market division was very visible. Currently, the Siemens market has emerged quite strongly from Europe and is strengthening in global markets to the detriment of Wonderware. **Wonderware is still the king of SCADA systems** due to the rich history and accustomations of many factories to this tool - according to the manufacturer's data there are over 200,000

MyScada

applications worldwide created with the help of different versions of Wonderware InTouch.

Other suppliers do not boast of such accurate data, however, we can assume that these are not small numbers due to the ubiquitous automation of the automotive, petro or metallurgy industries.

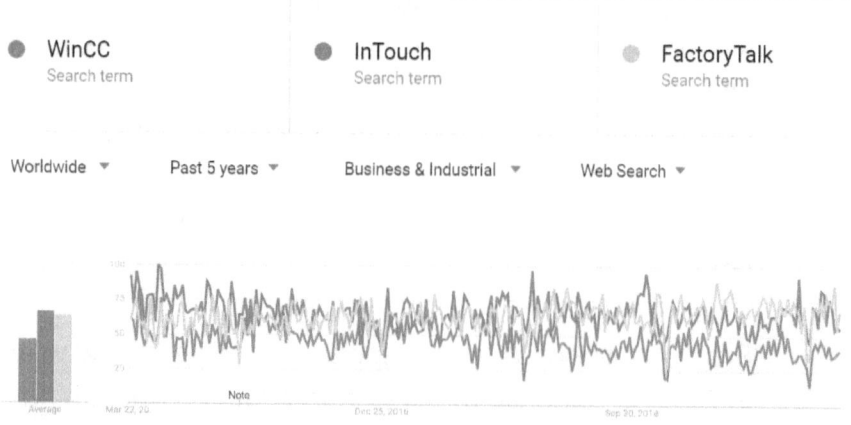

Foto: trends.google.com

● WinCC   ● InTouch   ● FactoryTalk

Foto: trends.google.com

MyScada

Siemens WinCC V7.x

Software description

SIMATIC WinCC is currently my favorite system for creating process visualizations (SCADA). WinCC (classic version e.g. V7.4, 7.5) is a scalable system that can be a simple stand alone project, an extensive server-client system based on redundant servers or an application created to work with many web clients. This approach allows it to be used in all types of industry and for all types of applications. One of the special features of WinCC is its total openness due to the possibility of connecting it with third-party software. As software for the most complex HMI

MyScada

tasks, WinCC is able to handle the largest projects with a lot of data.

Generally, I use WinCC as stand alone applications to manage one or more machines (usually between 2000 and 8192 external variables), although I have also recently created a server-client application with a redundancy option that controls 6 SEAT lines (> 40,000 variables). In the company where I work, I was involved in creating a multi-stand system where WinCC controlled over 30 devices (the entire production hall) along with connections between the SAP system.

Runtime and engineering software

WinCC Runtime is an integrated tool that allows broadly understood work with the system for the person designing the system, IT administrators, operators or maintenance department. The base of built-in controls and functionalities is very extensive and provides the programmer and the operator with a number of facilities while designing and working with the system. During operation, the system collects and archives data to the integrated MS SQL Server from which you can easily export and report data.

MyScada

WinCC CS is a collection of tools that meet all the requirements for performance and ease of use. Libraries and wizards make generating the project quick and easy, and also significantly reduces the possibility of error.

Features

- Faceplates technology
- Enables connection with most controllers and devices used in industry.
- OPC DA-client 2.05a/V3.0 in basic
- Each WinCC application can communicate without any restrictions over the network with other programs and applications.
- Is a container of ActiveX objects and .NET controls contained
- Has a built-in system of connection redundancy with devices, which ensures use in responsible systems.
- Support for multilingual applications
- High-performance data archiving (80 000 archive tags possible)
- Support for multi-monitor cards allows you to design graphically extended applications where a large one is required for the amount of information provided.

MyScada

- Application version compatibility is maintained - older ones can be converted to the newest ones. The work put into the execution of the system will be used in the future.
- Build in debugger for scripts (VBS, C)
- Virtualization - In addition to WinCC Clients, WinCC Server can be used with WinCC V7.x and later under VMware ESXi and with an engineering station under VMware Workstation, and Hyper-V from Microsoft. This means that complex client / server structures as well as single user systems can be configured through a virtual environment.
- Redundancy as option
- Process historian and Information server as option for web-based reporting
- WinCC/WebUX free for 1 user - view prepared system windows separated from normal work by device using HTML 5
- Integrated user administration and SIMATIC Logon that allow use chip card readers
- possible to automatically take over the AS messages of a S7-1500
- Licensing - basic software can be adapted to a large number of structures thanks to well-chosen licenses. The selected license can be easily upgraded based on the number of variables used.

MyScada

- Openness still progressing in basic technologies, operating systems, communication methods or the ability to integrate scripts, all on non-proprietary solutions
- Compliance with FDA guidelines - the basic system was designed as technologically independent. Nevertheless, it meets the specialized requirements of individual industries, e.g. FDA regulations for the pharmaceutical industry.
- WinCC ODK - possibility to interact with application from high level language programs
- Multitouch and gestures in visualization
- Build in VBA helps create SCADA by scripts and save a lot of time (and money)
- WinCC/IndustrialDataBridge - the bridge between industry and IT solutions

WinCC has a lot of packages which can extend basic functionality. Some of them are really expensive but give a lot of improvement without special programming work.

MyScada

Wonderware InTouch

Software description

Wonderware InTouch is a very advanced industrial software created for building visualization and process control. InTouch is also my first SCADA i learned almost 14 years ago in technical school. Most popular visualization package in the world. Wonderware InTouch software fully meets the guidelines for SCADA systems. It offers an easy-to-use and intuitive environment for designing applications and functionality that allows you to quickly design, test and implement valuable systems that provide users with data directly from control and production systems. InTouch is also an open and flexible software that allows you to adapt the application to current needs while maintaining a wide range of connections with

MyScada

devices and systems found in industry. Thanks to the concept of building applications from ready, easily configurable elements such as graphic objects, alarm state analysis objects or archiving and displaying process parameter history, the time and costs of starting the production visualization system, control and process analysis are reduced.

Runtime and engineering software

InTouch has a WindowViewer which is a built-in runtime tool. This tool is user-friendly due to the nice graphic design and the package of many amenities while working. Built-in trend or alarm controls meet the standards for scada systems, although for me 8 variables displayed on the trend is not sufficient. Fortunately, you can easily expand the InTouch system with a Historian which provides a much better approach to viewing archived data. WindowMaker is a programming tool for creating and editing visualization systems from screens, variables, alarms. The tool is neat and you can easily adapt the toolbars to our needs. Everything is close together and keyboard shortcuts work very well that speed up the creation of visualizations.

Personally, I created several dozen applications using Wonderware IntTuch from version 9.5 to version 2017 R2. It's nice to mention that the 9.5 application can be converted to the newest version in a few minutes without any major problems. The work put into the execution of the system will be used in the future.

Features

- ArchestrA technology.
- Easily integrates with the Wonderware System Platform, enabling Wonderware users to benefit from the latest solutions in the field of industrial applications, while retaining the benefits of investments already made.
- Enables connection with most controllers and devices used in industry.
- Can work as an OPC client, DDE client and server, and SuiteLink client and server at the same time.
- Each InTouch application can communicate without any restrictions over the network with other programs and applications.
- Is a container of ActiveX objects and .NET controls contained in new ArchestrA Graphics graphic objects, co additionally enriches the design

MyScada

environment when designing graphically and functionally extended applications.

- Has a built-in system of connection redundancy with devices, which ensures use in responsible systems.
- Support for multilingual applications
- Support for multi-monitor cards allows you to design graphically extended applications where a large one requires a large amount of information provided.
- Application version compatibility is maintained - older ones can be converted to the newest ones.

MyScada

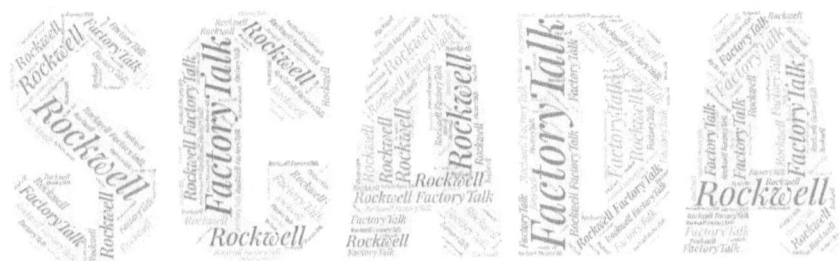

Rockwell FactoryTalk View Site Edition

Software description

FactoryTalk View Site Edition (SE) is an integrated software package for developing and running human-machine interface (HMI) applications that can involve multiple users and servers, distributed over a network. As a member of the FactoryTalk View family of products, FactoryTalk View SE provides all the tools you need to create powerful, dependable process monitoring

and supervisory control applications. In FactoryTalk View Studio, you can create network distributed, network station, or local station applications that mirror your plant or process. Use the editors in FactoryTalk View Studio to create and test the application components you need. Then set up the FactoryTalk View SE clients to let operators interact with the application after it is deployed. Can be scaled from an independent HMI system to a distributed visualization solution. FactoryTalk View Site Edition (SE) is a supervisory-level HMI software for monitoring and controlling distributed-server/multi-user applications.

Create complex applications that mirror the layout of a plant or process. A FactoryTalk View SE network distributed application can contain several servers running on multiple computers, connected over a network. Multiple client users can connect simultaneously to a network distributed application.

Runtime and engineering software

FactoryTalk View Studio is a tool in which we create, run but also carry out maintenance work on all components of the FactoryTalk View system. Built-in programming tools and editors

FactoryTalk View Studio is a FactoryTalk View design environment that provides the editors and tools needed to develop and test applications distributed across the network, network station and the human-machine interface of the local station (HMI). Contains editors for creating complete applications as well as client and server software for testing created applications. You can also use FactoryTalk View Studio to configure FactoryTalk security services for the applications you create

It is the least liked software for creating SCADA systems for me. I have the impression that it is very crude and is not adapted to current market trends which goes into openness, simplicity of implementation and connections with other applications. Nevertheless, the created scada systems, once created and running, still work today without any downtime or problems, which only confirms that FactoryTalk View is a world-class system to work as a SCADA class application.

Features

- Global objects, faceplates technology
- network applications can be opened and modified remotely

MyScada

- Easy access to tags defined in Logix family controllers
- Tagless HMI servers
- Possibility to separate system fragments into a separate application if it is not nested as local station applications
- System-wide diagnostics report and screen help
- Easy to determine operator troubleshooting help zone
- Health monitoring
- Redundancy feature
- Centralized authentication and authorization of system
- Support for multilingual applications (over 40 languages)
- Modify HMI tags at run time - all connected clients are updated without restarting the system
- Hot swap - HMI server can be replicated to the standby server.
- FactoryTalk Alarms and Events - complete alarm system
- Display only subscribed alarms(directly from Logix 5000 controllers)
- Dock and undock selected displays
- Extend the capabilities of FactoryTalk View SE with VBA

MyScada

- Interoperating with MS SQL Server and Microsoft Excel.
- Real-time and historical data trends (max 100 tags on trend control)
- Remote ODBC (Open Database Connectivity)
- Compliance with FDA guidelines - the basic system was designed as technologically independent. Nevertheless, it meets the specialized requirements of individual industries, e.g. FDA regulations for the pharmaceutical industry.

Other SCADA distributors

The scada market is huge and companies producing plc drivers, hmi panels also often have their own scada. You can not forget about dedicated systems written specifically for the solution, e.g. in C # or java.

Below is a list of systems that are recognizable in the industry:

- ICONICS - Genesis32 / Genesis64 SCADA system based on OPC-UA technology
- IC-VIEW - supervision and control systems for power installations and other media
- Ignition - SCADA and MES system based on Java technology. The software publisher is Inductive Automation.
- infoU - LSIS SCADA software
- ECS 2000 - Stange

MyScada

- National Instruments LabVIEW
- Pro Tool
- Proficy HMI / SCADA CIMPLICITY 11.0, by GE Digital
- Proficy HMI / SCADA iFIX (formerly GE iFIX), GE Digital, thanks to the use of network clients (e.g. Proficy Web HMI), visualization and control is available on mobile devices and via the Internet
- UniArt - international Israeli SCADA
- Vijeo CITECT - a system for supervision and visualization of industrial processes by Schneider Electric. In addition, it is also possible to archive data through the Vijeo HISTORIAN software
- WebHMI - SCADA / HMI system for the integration of automation systems, with the possibility of organizing access and management through browsers, with a WWW server, API and LUA scripting language.
- ZenOn HMI / SCADA software from COPA-DATA

I don't know which one to choose?

Choose the cheapest, the prettiest or the most interesting for you. Sounds silly or straight?

Imagine that it doesn't matter which SCADA system you get to know because the principle of operation is the same in all standard systems. I recommend choosing a known supplier who has a wide spectrum of modules and extensions because it will allow you to find more fun in getting to know the whole system. When deciding to learn or learn about the SCADA system, focus on getting as much information as possible (available online courses, YouTube and forum) where it will be fairly easy to ask about issues that bother us. Most of the systems offered on the market allow testing our system in the DEMO version (**e.g. 2 hours of work or messages informing about the demo version**).

Personally, I would choose WinCC or InTouch as the first learning system because the set of information and ready product demo are available on the web without fees and getting to know the advanced system will make it easier for us to switch to another supplier and after a few weeks of getting to know the environment we will be able to program.

MyScada

The work environment and "tricks" are the main differences between individual systems. You also can't forget about the errors and deficiencies that every system has its own. Personally, switching between systems repeatedly had to learn the current tricks in the new versions because each system upgrade added functionality and by the way can break an element that has always worked in some specific way, etc.

You can also go in the direction of high-level languages that in the age of the Internet, digitization is gaining a large fragment of the market. However, they are dedicated and building SCADA systems, e.g. in C # We build software from scratch and we never know if the total cost will be much higher than the solutions available on the market.

Good luck finding your favorite SCADA system!

SCADA - history or future?

Surveillance and data acquisition systems play an important role in automation projects. SCADA has consolidated its place as a necessity for automation systems. SCADA is also seen as an excellent tool for communication with any device via industrial protocols.

But what is the role of SCADA systems in the era of the Industrial Internet of Things (IIoT)? Will they still have a long life? How can SCADA evolve in the future?

Several SCADA platform providers are relinquishing their large protocol interface database to rely on third-party gateways with OPC UA or OPC DA instead. Tools such as the Intel IoT Gateway allow users to create an industrial gateway using free software and inexpensive hardware.

MyScada

Is SCADA dying?

The SCADA system can do many things that we don't have better tools for. Mainly it is about the alarm or data collection system which in the case of SCADA is very extensive. In addition, SCADA systems are great for local operations - strictly in the company, close to managed production elements where availability is most needed.

On the other hand, the IIOT era has come where data flow and capabilities are greater. Today's time likes technological innovations that use AI and Big Data sets the path for the development of SCADA systems in the near future. The most changed element compared to classic scada systems is the historical data collection system where instead of using SQL (static) databases, scalable cloud solutions are used. Keeping in mind the IOT4.0 approach where each element of the infrastructure connects and communicates with everyone we can notice that the SCADA system will soon have to connect to hundreds of thousands of points and not tens / hundreds as before.

SCADA changes this fact, but in my opinion it does not die. We still have a lot to do in terms of security and architecture.

Google trends analyze

Google trends is a great tool that allows you to track people's historical and current approach to typed phrases and keywords. If we enter the topic SCADA directly and set it to the global economy and look at the chart below, we have a great argument that SCADA is still at the TOP.

The world, people and the economy need information about systems, which allows us to conclude that the growing demand for digitization of markets is simply a great driving force to keep scada markets at a high level of interest.

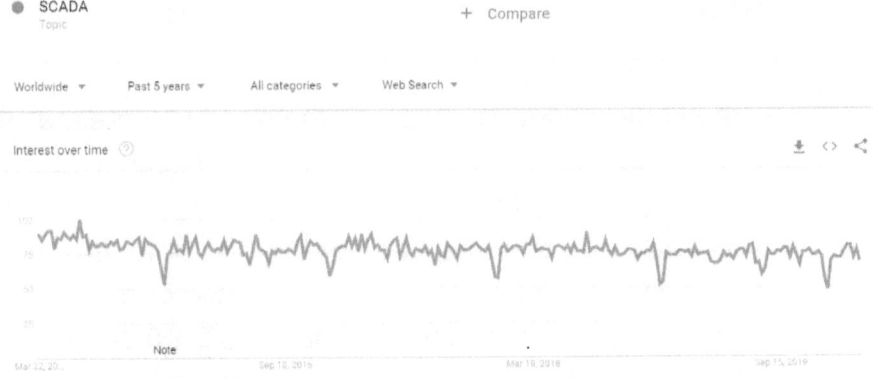

Foto: trends.google.com

SCADA world market

The scada industry (according to reports) grows on average a dozen or so percent annually. Which gives enormous opportunities for self-development to learn about interesting technologies
The budget is so huge that the age of specialists dealing with issues of scada Mes ERP systems is lacking. In the era of IOT4.0 there is not much data on cash flow in areas strictly associated with scada, but we can confidently assume that all activities aimed at automating the process also affect the development of scada markets.

SCADA engineer- modern superhero.

The scada market also carries the need to increase employment in the area of implementing such systems, servicing, implementation or maintenance. Activity in this area allows us as employees of this sector to develop our passions to travel the world, meet people or cultures. In addition, the current market trend to iOT 4.0 brings the industry market closer to the IT market, which translates into higher salaries below, I will try to present the conclusions of the search that I made at the beginning of

2020 based on several websites such as linkedin, payscale and indeed.

Let's look at the labor market in the pictures and searches below:

SCADA Engineer Salaries in the United States

Salary estimated from 132 employees, users, and past and present job advertisements on Indeed in the past 36 months. Last updated: February 14, 2020

Location

United States

Average salary

$99,745 per year

Salary Distribution

$32,000 $204,000

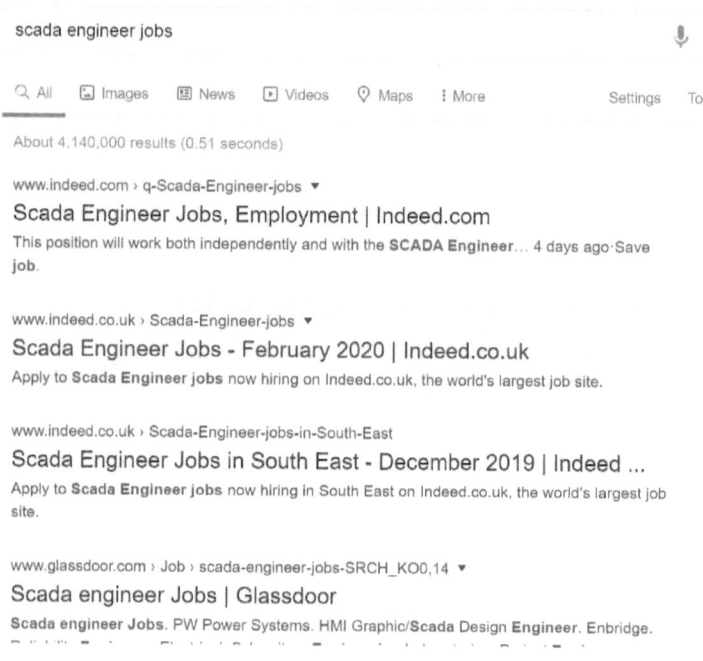

google.com

73

The labor market related to the creation, sale or maintenance of SCADA systems is huge. With 2-3 years of experience, we can earn really good money virtually anywhere in the world. If you are looking for a profession that will give you a chance for an interesting job and great opportunities, you should wonder whether it is worth investing in knowledge in the field of SCADA.

Maybe one day we will work in one team creating a system for automotive, aerospace or NASA!

Summary

Hi, this is the end of the basics about SCADA systems. I know that there may be elements that require clarification, some of the elements described may disagree or you have a different opinion. I understand and respect that. I am not a fiction writer with a light pen but a technical person who is easier to answer the problem than to describe the issues. However, I tried to take care of your time as much as I could. Let me know where I can improve the content of this book so that I can issue the second edition with corrections. I am also counting on suggestions on what to describe to describe additional chapters especially for you. Use the materials, leave a comment and see you.

PS. Remember you can find me on my blog, which is starting soon. Go ahead and join the mailing list now - we'll keep in touch!
https://myscadaworld.com/

PS2. Remember"You can't repair the world with just one SCADA !"

MyScada

References

1. https://en.wikipedia.org/wiki/SCADA
2. http://myscadaworld.com/
3. https://www.wonderware.com/
4. https://support.industry.siemens.com/cs/mdm/109760739?c=1
   09153147019&lc=en-WW
5. https://support.industry.siemens.com/
6. https://www.automation.siemens.com/salesmaterial-as/brochu
   re/en/df_fa_i10077-00-7600_ipdf_wincc_systemoverview_en.
   pdf
7. https://www.youtube.com/watch?v=Xymhe-0neN4
8. https://literature.rockwellautomation.com/idc/groups/literature/
   documents/pp/ftalk-pp013_-en-p.pdf
9. https://literature.rockwellautomation.com/idc/groups/literature/
   documents/um/viewse-um006_-en-e.pdf
10. https://www.indeed.com/salaries/scada-engineer-Salaries
11. https://trends.google.com/trends/explore?q=SCADA
12. https://trends.google.com/trends/explore?cat=12&date=today
    %205-y&q=WinCC,InTouch,FactoryTalk
13. https://www.payscale.com/research/US/Job=SCADA_System
    s_Engineer/Salary
14. https://www.marketsandmarkets.com/Market-Reports/scada-m
    arket-19487518.html
15. http://www.picmet.org/new/conferences/17/handouts/Keynote
    _Theis.pdf
16. https://trends.google.pl/trends/explore?cat=12&date=today%2
    05-y&q=WinCC,InTouch,FactoryTalk

MyScada